Spirit of Flight

POEMS OF AVIATION COMPILED BY IAN GENTLE

NMS Publishing

Published by NMS Publishing Limited, Royal
Museum, Chambers Street, Edinburgh EH1 1JF

Collection © NMS Publishing Limited 1999
Poems © individual authors

Information about our books is available at
www.nms.ac.uk

**British Library Cataloguing in Publication
Data**
A catalogue record of this book is available from
the British Library

ISBN 1 901663 16 7

Designed by NMS Publishing
Printed in the United Kingdom by Cambridge
University Press, Printing Division

Cover illustration: *Flycatcher II* by Keith Maddison
G.Av.A, Winner of the Arthur Gibson Award for
the Aviation Painting of the Year 1997.

Contents

INTRODUCTION

The spirit of flight. What is it? How did people conceive the dream of flight before it became a reality? Then there is the experience of flight itself, observed and recorded by people in the twentieth century. The combination of concept and experience make up what I consider to be the spirit of flight.

Poets have been inspired by these concepts and experiences, and their variety is what this anthology is about. Aviation in war and peace has fed the poetic imagination, and these aspects are dealt with here. Mythological references abound, as in Ken Barris' poem *A question of height*, which, through compelling use of metaphor explains the lesson of Icarus' flight. The technology of war is criticised - from a technological point of view - in the anonymous *That Thing*, written by an unknown Fleet Air Arm officer. Specific events and individual aircraft are celebrated in verse ranging from *Tangmere 1940* by Angus Duncain to Willie Hershaw's *Lockerbie Elegy*.

Much of the writing is celebratory although the shadow side of flying, be it expressed as the banality of package travel or the horror of aerial bombardment, is not ignored.

Spirit of Flight was inspired by the historic collections at the Museum of Flight, East Fortune. It is appropriate that the Museum – the word is derived from the Greek for 'temple of the Muses' – has been the impetus for this collection of poetry.

Here is poetry for all ages and every mood, in works that prove that the human imagination can soar above the earth and reach towards the eternal.

I wish to thank everyone who helped with the production of *Spirit of Flight*, starting with Maureen Kerr who did the word processing. I also wish to thank Janis Kemp and June McDonald for their preliminary work. Helen Kemp is to be thanked for editing the final text, and Elizabeth Robertson for the design. Gavin Sprott conceived the idea and Adam Smith took it on board. The staff of the Scottish Poetry Library, Edinburgh, particularly Tessa Ransford, are to be thanked for helping find the sources for me. I also wish to thank the staff of the National Library of Scotland, Edinburgh and the Mitchell Library in Glasgow.

I am most grateful to the poets and publishers who allowed their work to be reproduced here, and a very special thanks must go to you, the reader.

The publishers are grateful for the following permissions: Bloodaxe Books for Fred D'Aguiar *Flying to Nowhere* from *British Subjects* 1993, Andrew Greig and Kathleen Jamie *A Flame in Your Heart* 1986, Josef Hanzlík *Air Show* from *Selected Poems* 1993, Miroslav Holub *On Daedalus* from *Poems Before and After* 1990, Lawrence Sail *Reichert's Leap* and *Flight 000* from *Out of Land: New & Selected Poems* 1992, Pauline Stainer *Skydivers* from *Sighting the Slave Ship* 1992, Martin Stokes *Hubris, off the white cliffs* from *The First Death of Venice* 1987; Ken Barris and Snailpress for *A question of height* from *An advertisement for air* 1993; *Airmen from Overseas* from *The North Star and other poems* 1941 by permission of The Society of Authors on behalf of the estate of Laurence Binyon; Carcanet for Paolo Buzzi *Highway to the Stars* and Filippo Tommaso Marinetti *The Futurist Aviator* from *The Blue Moustache* 1981; Scottish Cultural Press for Gerry Cambridge *North* from *The Shell House* 1995, Willie Hershaw *Lockerbie Elegy* from *The Cowdenbeath Man* 1997, Anne Macleod *Icarus* from *Standing by Thistles* 1997; Curbstone Press for Ernesto Cardenal *Flight over the*

Homeland without Stopover from *Flights of Victory* 1988; Angus Duncain for *Tangmere 1940* from *Reflections in the Millpond* 1987; Gus Ferguson for *Businessmen in Flight* from *Carpe Diem: Poems & Drawings* 1992; Francis Gallagher for *Lockerbie, the easy war* from *Shadow Song* 1993; New Directions for Lars Gustaffsson *The Wright Brothers look for Kitty Hawk* from *The Stillness of the World before Bach*; Tony Harrison for *Listening to Sirens* and *The Act* from *The Gaze of the Gorgon*, Bloodaxe 1992; James Inglis for *The DH Dragon Rapide*; Polygon for David Kinloch *Icarus 2* from *Paris-Forfar* 1994; James Kirkup for *In a Sailplane* from *The Prodigal Son, Poems 1956-1959*, Oxford University Press 1959; Douglas Lochhead for *In memory of James Eagleson* from *Collected Poems: The Full Furnace* 1975; Christine De Luca for *Airborne over Orkney* from *Voes and Sounds*, Shetland Library 1995; Random House for Norman MacCaig *Daedalus* from *The Collected Poems*, Hogarth Press 1990, Vernon Scannell *...Jet* and *Flight of balloons* from *On your cycle, Michael*, Red Fox 1991; Peter M McCulloch for *The Flying Visitor* and *The Harvard*; University of Queensland Press for Roger McDonald *Airship* from *Airship* 1975; William Meredith for *Homeric Simile* from *Earth Walk: New and Selected Poems* 1970; William Neill for *Prestwick Airport* from *Just Sonnets* 1996; HarperCollins for Brian Patten *Tina's Flight* from *Armada* 1996; Enitharmon Press for Jeremy Reed *Throttling the pilot* from *Saints & Psychotics* 1979; James Reid for Fly; Dionysia Press for Susanna Roxman *Ballad of a Balloonist* from *Broken Angels* 1996; Gavin Sprott for *Planes*; Daniel Stolfi and Snailpress for *In Praise of Flying* from *Vectors: Selected Poems 1983-1991* 1992: John Updike *Airport* from *Collected Poems 1953-1993* 1993 by permission of Alfred A Knopf Inc.

Every effort has been made to trace copyright holders. If we have made errors or omissions we would be grateful to hear from the authors or publishers concerned.

Flying to Nowhere

A four-seater airplane I'm in drops low
over the Thames Barrier, moved to some
jungle-type setting, by which I mean
transplanted as places tend to be in dreams.

Jungle or not there's a cathedral - don't ask how -
in the middle of it, guarded by an army
you'd be foolish to meddle with;
dripping sophisticated weaponry, muddy from

a recent successful campaign, they ride in
to a town duty-bound to celebrate their return
in a mile's parade or capitulate. We (I have
company, three others, all destined for that plane)

run through undergrowth high as a garage, in a big
effort to stay camouflaged, reach the plane
I decide not to fly, acting as co-pilot instead.
This is when we land near the Barrier.

This is where we play catch with various sized balls,
in a green field that contains our mightiest throws,
green with care. This is how I come to call it paradise
and no one argues over the name and it sticks.

We buy our provisions in a nearby store
where the one language is a French patois
delivered with clucks on the upper palate
by the tongue, where the currency is English pounds.

My girlfriend cuts her thumb cutting a joint of ham.
I ask her to stand in a corner with her hand
in the air while I get through our shopping list.
There's no blood to speak of, but it stings all the way

back to the haven I'll call for argument's sake
heaven, since it's paradise in the sense
that we're all of us dead; we don't know
or we know and don't care; both feel the same.

Fred D'Aguiar

THAT THING

Why should the unoffending sky,
Be tainted and corrupted by
This product of a twisted brain,
That's aeronautically insane,
This vile and hideous abortion,
Devoid of beauty and proportion,
That people call a Barracuda,
Whose form is infinitely cruder
Than any other scheme or plan
As yet conceived by mind of man.
To see it stagger into space
Would bring a blush upon the face,
Of the most hardened Pharisee
Within the aircraft industry.
But I suggest we don't decry
This winged horror of the sky;

But keep it 'til the War is won,
And then we'll all join in the fun.
Festoon the wings with fairy lights
And wheel it out on gala nights,
Thus so we'll help dispel the rumour
That Britons have no sense of humour.

Anon FAA Officer

A QUESTION OF HEIGHT

For the modern understanding, Icarus' problem
lay not in his father's aerial design
but in the materials of his wings.
Not wax but compound polymer; the feathers
could have been aluminium or carbon fibre,
even shellacked canvas at a pinch
stretched over balsa wood, even acres of silk
like a roped-in cloud billowing behind.

As to soaring too close to the sun, he lacked
ground control. This is a position many
today understand well, choosing never
to leave the labyrinth of analysis
and other forms of self-perpetuating
enquiry, sensing the terror of too great
a height, the giant tectonic plates of air
and the unmitigated, the cruelly
resolving streams of light.

Ken Barris

Airmen from Overseas

Who are these that come from the ends of the oceans,
Coming as the swallows come out of the South
In the glory of Spring? They are come among us
With purpose in the eyes, with a smile on the mouth.

These are they who have left the familiar faces,
Sights, sounds and scents of familiar land,
Taking no care for security promised aforetime,
Sweetness of home and the future hope had planned.

A lode-star drew them: Britain, standing alone
Clear in the darkness, not to be overcome,
Though the huge masses of hate are hurled against her.–
Wherever the spirit of freedom breathes, is Home.

Soon are they joined with incomparable comrades,
Britain's flower, Britain's pride,
Against all odds despising the boastful Terror;
On joyous wings in the ways of the wind they ride.

From afar they battle for our ancient island,
Soaring and pouncing, masters of the skies,
They are heard in the night by the lands betrayed and captive
And a throbbing of hope to their thunder-throb replies.

To dare incredible things, from the ends of ocean
They are coming and coming over the perilous seas.
How shall we hail them? Truly there are no words
And no song worthy of these.

Laurence Binyon

Highway to the Stars

We were flying at a hundred miles an hour.
The night advanced a thousand miles a minute.
Whorls of lead, clouds shifted in the sky;
the mountains reared their peaks of anthracite.
The lake below us looked like a basin filled
with the ink of cuttle-fish.
Lights began to flicker along the shore.
On and on we sped, as through a mighty sea,
cross-hatched with lines and shadows,
our nostrils flaring in the windy mist.
The landscape rumbled to our engine's roar,
and my heart sang an ode impossible to repeat.
Behind us the women chattered gaily, like fountains,
their veils caught in the blow, and trailing perfume.
Suddenly, lights flaring –
we streamed away, a comet with a tail,
on a highway of bright abysses of stone and water and air
fixed in the devil-prism of our beam.
The rain poured, spilling diamonds.
Blackness all about,
the women's eyes shot sparks,
and in that wide and glowing orbit
we lived through a grand vertigo of whirlpools
like strange creatures of fire
from another part of the far-flung galaxy.

Later,
bone-drunk on starry metaphysics,
how like gods - in the stillness of the spirit's cloister -
did we warm our flanks and our dreams
by the glowing meteors in the fireplace!

Paolo Buzzi
translated by Felix Stefanile

NORTH

The jewel-lit noon of the bombing,
 In a sparse land of stone
I walk the dazzling shore
 To Faraid Head, alone.

Mountains sleep on the skyline
 In the silver and emerald day,
And there's the target island,
 Tiny, miles across the bay.

I leave the shore: a short cut
 Through silent dunes and grass,
To reach the closest vantage,
 And watch all that may pass:

When over my head, three jets streak –
 Silence, then a roar.
The horizon's hidden here,
 By hillocks all before.

So I run, run, run up one –
 There's the island suddenly
And the first jet shrinking over it
 On a white cloud, darkly.

I raise binoculars tremblingly:
 Poise in a black-ringed bubble of light
The island screed from earlier bombs,
 The target hut on its extreme right...

And see in the bubble how delicately
   The jet-loosed bomb twirls down,
Like a tiny seed that, spinning,
   Mirror-flashes to the sun.

And glittering down through deep air —
   It kisses the island, flashes —
Silent, orange on white and blue
   Outblossoms an instant, vanishes.

I lower binoculars tremblingly:
   Unchanged the island sits there still,
Smoke-columned now in brilliant air,
   As I here, on my small hill

Listen long for the shock —
   As clouds like bubble-pearls stand round,
As the smoke rises, wisping —
   For the bomb has made no sound.

Yes, listen in eerie silence here,
   And with a feigned desire
Think this less than it appears,
   Merely a silent fire.

Think it less as seconds tick
   And over the miles of water
News now glides of a cataclysm
   Not meant for slaughter —

Then the tight orgasmic bang.
   As if the sky should crack, fall,
And mountains, crofts, blow away,
   Though all stands just as usual —

Except Earth jolts, as if a giant
   Jumped to ground, off the sun;
Yelping gulls upflurry like snow,
   A flock of far sheep run;

And horizons to sharpness tremble back
   As the second jet roars in,
Like a thought catapulted over the island,
   And each wing like a fin.

And six dream-times all happens
   As massive Earth turns, slow:
Flash – calm – bang-rocked shining world –
   Till the jets, in seconds, go.

And dazzle spreads over the wrinkled sea
   As an oystercatcher, nesting, calls;
As in a waiting amphitheatre,
   Breeze-ruffled silence falls.

Faraid Head, Durness, Sutherland, 1986

Note: RAF target bombing of An-garbh-eilean (stony island) off Cape Wrath,
can be watched from the nearby headland of Faraid Head. All details in the poem
are documentary.

Gerry Cambridge

Flight over the Homeland
without Stopover

First came the lake, calm.
         And in it
my place, what was my home, Solentiname.
         All the islands grouped together: they seemed like just one.
But I could distinguish them one by one from afar, say their names.
The point where our community was. Everything demolished.
The library burned. That hammock under a roof of palms
         with the lake in front.
Elvis and his guitar. Where did the guardsmen bury them?
         Further on, La Zanata, solitary in the middle of the lake.
Where we went to fish for *guapotes*.
         Almost right below the plane but unreachable.
I would now be seeing this plane from the fibre-thread hammock.
The stewardess offers us drinks. I ask for a whisky on the rocks.
I drink it my view glued to the window which separates me.
No one on the plane suspects this tragedy.
         The exile looking at his homeland.
There behind the window.
         But between me and my land there is an abyss.
My land at war. It was after the events of September.
I was flying to a congress of the Socialist International in Lisbon
to cry out for the blue beloved that I was seeing from the clouds.
I want to see the places of the large demonstrations. The sites
of the meetings by the light of campfires at night,
the shots at armoured cars patrolling the streets, the masked children,
the contact bombs, where they move in the brush,
the mothers crying, Monimbó, where they throw the assassinated.
I want...

That for which it is beautiful to die, all blue and clouds.
The beloved geography denied to me.
I didn't see cities. Only blue mountains.
                              And suddenly
        I saw Estelí,
to my unhappiness, I knew that this was Estelí:
        A black and murky quadrilateral
among green fields.
Not the whites and motley colours of tiny houses:
            but a stain the colour of coal and ash
like a burnt up corpse.
No one else has seen a thing ... And the stewardesses
start serving plastic food as if nothing's happened.

Ernesto Cardenal
translated by Marc Zimmerman

THERE ARE NO FRONTIERS IN THE SKY

There are no frontiers in the sky,
The clouds exact
No custom dues; the long, bright lanes of sun
Ask not the parentage of those who fly.

What care the winds the colour of a face
Or does the night
Play jealous with her stars unless she knows
The status of an airman's dwelling-place?

This infinite upon whose little rim
Man dares to crawl
Assesses not his politics or creed
But with indifference slays or succours him.

O C Chave

Native Land; Spitfire Pilot; England, 1940

And entry's a leather-worn gesture,
I'm into the cockpit,
Pram-fitted, firmed in.
Outside me, is summer
In layers,
A ransom of earth;
With the greenest of skins.

K F Daly

Tangmere - 1940

Morning mists clearing
Telephone ringing
Engines roaring
Young men running

Fighters soaring
Vapour trailing
Eagles duelling
Skidding, Swirling

Circles tightening
Closing, closing
Mirror glancing
Safety catch lifting

Gunsight filling
Button pressing
Fire spitting
Scoring, Hitting

Sudden thudding
Right arm stinging
Gauges rising
Control slipping

Engine screaming
Coolant streaming
Smoke filling
Hood sliding

Air rushing
Falling, falling
Ripcord pulling
Jerking, braking

Floating, floating
Right arm soaking
Corn stooks ripening
Gratefully landing.

Angus Duncain

BUSINESSMEN IN FLIGHT

In motion thirty thousand feet above
The ground, like sardines in a streamlined tin,
We arc the earth with pinion and fin
Intent on missions mercantile. The love
Of flight, the sheer suspense of disbelief,
The graceful gravity-defeating act
(An arabesque of faith), does not, in fact,
Excite the pulse, cause ecstasy or grief.
We who travel, but not for travel's sake,
Are flung aloft by market forces, still,
Take pride, while cloistered in an aerobus,
That Marco Polo, Mungo Park and Drake,
Explorers who escaped the common mill,
All represented merchants. (Just like us!)

Gus Ferguson

LOCKERBIE, THE EASY WAR

they are arrivals that will never be made now
and departures that will go on for ever
the relatives will look up stupidly tomorrow
for their loved ones to come in then remember

they were all sacrificed to some impossible cause
for terror is the last throw of socialism
the Aztec god that demands blood as release
and revenge for how final its own defeat is

on flight 103, the 259 victims of history
complete innocents caught in the brutal crossfire
of fanatics whose whole way of life is arbitrary
slaughter the injustice of death and evil's laughter

German police playing their own game lost the way
to hide their incompetence let the terrorists
slip through the net the purpose of security
is to conceal your own mistakes at any costs

mad Gadaffi cranks his socialist revolution
the old man of the mountain in his terror state
sends out assassins to murder women and children
another moslem victory the intolerant religion of death & hate

and after the numbness of rage the self anger
loss of life blisters the waste of opportunity
all that putting off telling showing the other
how much you care as if you had eternity

and now you would sell your soul to recall
the dead even for a moment to tell them
all that's in your heart hardest of all
you'll both die without saying you love them

the crimes of history require the participation of too many
and for that reason no one is punished no one is guilty

Francis Gallagher

From A FLAME IN YOUR HEART

One by one they will return,
throttling down over perimeter wires
of remembered airfields, then taxi up
to abandoned huts.

They look around in the rain.
There are pigs in the Control Tower,
in a shelter one finds an old 'Picture Post',
a French letter between the pages.
They shrug and laugh, the youngest
bites his lip.

They share cigarettes, and talk
for an hour by the Dispersal Hut,
then one by one take off and climb
above the clouds, where it is always blue,
burning and burning at that summer's end.

Andrew Greig and Kathleen Jamie

The Wright Brothers look for Kitty Hawk

In an agitated dream I saw everything explained:

Otto Lilienthal soar impressively in his glider
down the steep hill in Grosslichterfelde.

A fierce wind blew, which drove kites,
and someone monotonously talked about "the gnostic darkness."

It was a warning, a whisper that came and went.

Bakunin steps aboard the cargo-steamer Andrew Steer,
one spring day, in the harbour of Nikolaevsk, among sheds and
shops.

In the 19th century the sea often smells musty in a calm.
The revolutions are being prepared. Seafire sparkles.

And Milton Wright, bishop in The Church of the United Brethern,
presents his sons Wilbur and Orville with one of Pénaud's models:

not unlike a misformed bird with a hungry neck.
Wind-tunnel experiments at the cycle factory's range

and the dry sand which smokes in stubborn wind.
What is bad or good about a kite? It flutters,

rises in an abrupt stagger but with a dead motion
at the moment when it should tear the thread

the much too short thread. In Africa the locomotives rust,
and the steamship "Savannah" with fluttering streamers

over the unreal blue sea. Solemn smoke.
Nature is always obvious: the straddle-helm and the propellor.

Dresden. Hanoi. And "the gnostic darkness."

Lars Gustafsson
translated by Robin Fulton

From AIR SHOW

This year's air show at Letňany airfield
was held in bad weather
A strong wind blowing and many showers
from a grey sky
Among the spectators lovers
hugged happily under umbrellas
designers of light summer dresses
were in despair
But I admired
the incredible aerobatic game
of jet fighters in threes or sixes
soaring and diving
in a bizarre minuet or polonaise.
When
they flew off the television screen
I ran to the balcony
and watched them in the real sky
returning home
Already beyond range of the cameras
but still with the same symmetry
For breaking it
would mean death

When it was over
I turned off the TV
in my flat close to the airport
and glanced
at the postcard
tucked under the glass top of my desk
Pilot Alphonse Pégoud
standing there in a leather jacket and riding boots
in front of his monoplane Blériot

Though French his moustache
strangely reminded you of Kaiser Wilhelm
He looks grave right leg slightly bent
goggles over his peak-cap

The postcard was issued by the Prague Ladies' Section
of the Central School Foundation
and bears the inscription
Pégoud's take-off in Prague
25 and 26 December 1913

Letňany airfield didn't exist then
and had no air shows
Pégoud hovered over the city's Letná plateau -
to Prague onlookers his stunts were miraculous
He not only dived
and rolled but also
was the first in the world
stunningly to loop the loop

He flew in a famous but long outdated type
or rather
in an outdated but famous one
Because in the same monoplane
already in 1909
Louis Blériot had crossed the Channel

Pégoud stands in front of his machine
which looks as vulnerable
as our children's models of long ago
made of plywood and parchment paper
His plane's front wheels
must have come from a bicycle
The engine is half exposed
the canvas fuselage held together by thin wires

Everything is as light as possible
as if Alphonse Pégoud were to become
airborne on his own

He's not smiling
his lips aren't parted like a film star's
he doesn't smoke a cigar
into the lens of the folding hand camera
he trusts himself
believes he'll put on a good show

Only he has no idea
that in a year there will be war
and he'll be flying over the front lines
without loops without applause

Only he doesn't know
that in two years
he'll be shot down
during a reconnaissance flight

That he is going to die in the debris
of a French air force biplane
bones crushed
and hair aflame

That he is going to die
just like his brilliant admirer
Antoine de Saint-Exupéry
twenty-nine years later

During a combat flight

Yes during a combat flight
The author of looping
and author
of *The Little Prince*

Because war likes to murder
those we loved in childhood . . .

Josef Hanzlík
translated by Jarmila and Ian Milner

LISTENING TO SIRENS

Was it the air-raids that I once lived through
listening to sirens, then the bombers' drone
that makes the spring night charter to Corfu
wake me at 2, alarmed, alert, alone?
I watch its red light join the clustered stars
in the one bright clearing in the overcast
then plummet to become a braking car's
cornering deserted side-streets far too fast.

My lilac purples as the headlamps pass
and waft it in, that same lilac smell
that once was used to sweeten mustard gas
and induce men to inhale the fumes of hell.
A thin man from that War who lived round here
used to go berserk on nights like these,
cower, scream, and crap his pants with fear
whenever he scented lilac on the breeze.

Senses that have been blighted in this way
or dulled by dark winter long for the warm South,
some place we hollow out for holiday,
and nothing spoils the white wine in the mouth.
I drag my senses back into the dark
and think of those pale Geordies on their flight.
I'll still be oblivious when they disembark
dazzled by the blue and the bright light.

Tony Harrison

25

THE ACT

(for Michael Longley & James Simmons)

Newcastle Airport and scarcely 7 a.m.
yet they foot the white line out towards the plane
still reeling (or as if) from last night's FED
or macho marathons in someone's bed.
They scorn the breakfast croissants and drink beer,
and who am I to censure or condemn?
I know oblivion's a balm for man's poor brain
and once roistered in male packs as bad as them.
These brews stoke their bravado, numb their fear
but anaesthetise all joy along with pain.

To show they had a weekend cunt or two
they walk as if they'd shagged the whole world stiff.
The squaddies' favourite and much-bandied words
for describing what they'd done on leave to birds
as if it were pub-brawl or DIY
seem to be, I quote, 'bang', 'bash' or 'screw',
if they did anything (a biggish if!)
more than the banter boomed now at the crew
as our plane levels off in a blue sky
along with half-scared cracks on catching syph.

They've lit Full Strengths on DA141
despite NO SMOKING signs and cabin crew's
polite requests; they want to disobey
because they bow to orders every day.
The soldiers travel pretty light and free
as if they left Newcastle for the sun,
in winter with bare arms that show tattoos.
The stewardesses clearly hate this run,
the squaddies' continuous crude repartee
and constant button-pushing for more booze.

I've heard the same crude words and smutty cracks
and seen the same lads on excursion trains
going back via ferry from Stranraer
queuing at breakfast at the BR bar,
cleaning it out of *Tartan* and *Brown Ale*.
With numbered kitbags piled on luggage racks
just after breakfast bombed out of their brains,
they balance their empty cans in wobbly stacks.
An old woman, with indulgence for things male,
smiles at them and says: 'They're nobbut wains!'

Kids, mostly cocky Geordies and rough Jocks
with voices coming straight out of their boots,
the voices heard in newsreels about coal
or dockers newly dumped onto the dole
after which the army's the next stop.
One who's breakfasted on *Brown Ale* cocks
a nail-bitten, nicotined right thumb, and shoots
with loud saliva salvos a red fox
parting the clean green blades of some new crop
planted by farm families with old roots.

A card! The stewardesses almost throw it
into our laps not wanting to come near
to groping soldiers. We write each fact
we're required to enter by 'The Act':
profession; place of birth; purpose of visit.
The rowdy squaddy, though he doesn't know it
(and if he did he'd brand the freak as 'queer')
is sitting next to one who enters 'poet'
where he puts 'Forces'. But what is it?
My purpose? His? What are we doing here?

Being a photographer seems bad enough.
God knows the catcalls that a poet would get!
Newcastle-bound for leave the soldiers rag
the press photographer about his bag
and call him Gert or Daisy, and all laugh.
They shout at him in accents they'd dub 'pouf'
Yoo hoo, hinny! Like your handbag, pet!
Though what he's snapped has made him just as tough
and his handbag hardware could well photograph
these laughing features when they're cold and set.

I don't like the thought of these lads manning blocks
but saw them as you drove me to my flight,
now khakied up, not kaylied but alert,
their minds on something else than *Scotch* or skirt,
their elbows bending now to cradle guns.
The road's through deep green fields and wheeling flocks
of lapwings soaring, not the sort of sight
the sentry looks for in his narrow box.
'Cursed be dullards whom no cannon stuns'
I quote. They won't read what we three write.

They occupy NO SMOKING seats and smoke,
commandos free a few days from command
which cries for licence and I watch them cram
anything boozable, *Brown Ale* to *Babycham*
into their hardened innards, and they drain
whisky/lemonade, *Bacardi/Coke*,
double after double, one in either hand,
boys' drinks spirit-spiked for the real *bloke*!
Neither passengers nor cabin crew complain
as the squaddies keep on smoking as we land.

And as the morning Belfast plane descends
on Newcastle and one soldier looks,
with tears, on what he greets as 'Geordie grass'
and rakes the airport terrace for 'wor lass'
and another hollers to his noisy mates
he's going to have before their short leave ends
'firkins of fucking FED, fantastic fucks!'
I wish for you, my Ulster poet friends,
pleasures with no rough strife, no iron gates,
and letter boxes wide enough for books.

Tony Harrison

LOCKERBIE ELEGY

The frown of his face
Before me, the hurtle of hell
Behind, where, where was a, where was a place?
    The Wreck Of The Deutschland – *Gerard Manley Hopkins.*

December is dark and fas
Mirk ower Border march and moor.
December is a shroud. The banshee blaws
Thru the langest day and darkest hour.
The year turns like a hearse wheel, sweir
Ti win the slottry world awo fae deith,
Ti rax a resurrection, speir
Despair and horror brak wi the Sun's bricht breith.

Our ancestors fired folk,
They brunt the weerded anes as sacrifice,
Handselled the dark wi flesh and smoke
That the laich licht micht kyth.
As the reek rose - the bodies thirled ti a wheel
O fire-flaucht-the watchers witnessed waste
O fleshly form. This fire-festival
Loused Pagan spirits ti wrocht a future hairst.

Us, we lauch and snirt
At them, screevinless, withoot science or skeill.
Smug-faced, in livin-rooms we sit
As if we birled Fate's fell wheel.
We gawp at graves on screen, our conscience free
Ti mak nae connection wi the deith-tales tellt.
Gorbachev and Bush like dark angels flee
Abune our heids. Horrors are speired, are heard but no felt.

Until the day o sky-doom,
The lift-lapse when the carry crashes,
When Deith is biddan til the livin-room,
When the roof-tree rends and smashes,
When the yird-bound plane plummets fae space
When the news is torn bluidy fae the broken screen,
Unbelief screeved on its face -
The crushed corpse on the carpet screeved for ay on our een.

The sky-wrack strewn
Wi torn and twisted bodies scattered,
The wing-wreck ligs ower field and toun,
Metal-mangled and bomb-battered.
The horror o this fire-ship, fresh-fa'n,
Wad seecken the mind speired in its ugsome hale:
Lockerbie wauks ti a hellish dawn,
Its crater-carved streets unmade til Paschendale.

Pray, pity them then,
That fell in fear fae unfathomed hecht,
Fell in a howlin gyre, forlane,
In a yowlin gale that blew that nicht.
Did they fa unkennin til Deith's dark deep
Fae the harrowin hertbrak that hell brocht?
God gie their screamin sowels sleep,
Pray, pity them as they thocht their last thochts.

Wha farrant their foul fate?
Wha doomed them ti dee? Wha schemed the ploy?
This scunnerfu sacrifice – this deith-gate,
Brocht nae pagan life-joy.
A deed ti mak maist blanch and blate –
Whit cause, whit clan wad ca such dool? Whit men,
Whit wrang worth such a hate?
Whit leid a life when life is a' we hae and ken?

The howlin engine banshee
Pibrochs the plane – mocks mad Man's will.
Born ti be a burnin ba o energy
Birlin thru the universe for guid or ill
We choose ti destroy, murder, maim and unmak
Afore we ken a peerie piece o our place,
O our circumstance, let the cruel cloods crak!
That we in madness micht speir the froon on God's face.

We clatter doon til end-shroud
As Vulcan, Daedalus and Lucifer fell.
We bide on the yird-brek thru brief life-clood
Ti non-being, spirit-louse or hell.
Ane day we maun dree their weerd – the wecht
O Life will drag us ablo tae, deaved wi pain,
Hard doon wi a horror o hecht –
Pray, pity us that fa as December deith-rain.

William Hershaw

31

ON DAEDALUS

Daedalus potters in the labyrinth.
Self-generating walls.
There's no escape.
Except wings.

And all round – those Icaruses. Swarms of
them.
In the towns, in the plains, on the uplands.
In the airport lounge (automatic
goodbyes);
at the space control centre (transistorised
metempsychosis);
on the sports ground (enrolment of pupils
born 1970);
in the museum (blond seepage
of beards);
on the ceiling (a rainbow stain
of imagination);
in the swamps (hooting of night,
born 1640);
in the stone (Pleistocene finger
pointing upward).

Time full of Icaruses,
air full of Icaruses,
spirit full of Icaruses.

Ten billion Icaruses
minus one.

And that even before
Daedalus invented
those wings.

Miroslav Holub
translated by Edwald Osers

THE DH DRAGON RAPIDE

Caressing the air,
A song in the wind,
Airborne at last,
Fulfilling a dream,
Fast darting about,
In my Dragon Rapide,
Swiftly aloft,
With twin Gipsy Queens.

James Inglis

ICARUS 2
after Robert Desnos

Somewhere Eskimos are weeping
For failing to bury enough sunlight
In their igloos.
But in Outer-Siberia a man with a bear-skin
Telescope laughs like the steppe,
Detecting the changing smiles of Icarus
At different pitches of the air.

David Kinloch

IN A SAILPLANE

Still as a bird
Transfixed in flight
We shiver and flow
Into leagues of light.

Rising and turning
Without a sound
As summer lifts us
Off the ground.

The sky's deep bell
Of glass rings down.
We slip in a sea
That cannot drown.

We kick the wide
Horizon's blues
Like a cluttering hoop
From round our shoes.

This easy 'plane
So quietly speaks,
Like a tree it sighs
In silvery shrieks.

Neatly we soar
Through a roaring cloud:
Its caverns of snow
Are dark and loud.

Into banks of sun
Above the drifts
Of quilted cloud
Our stillness shifts.

Here no curious
Bird comes near.
We float alone
In a snowman's sphere.

Higher than spires
Where breath is rare
We beat the shires
Of racing air.

Up the cliff
Of sheer no-place
We swarm a rope
That swings on space.

Breezed by a star's
Protracted stare
We watch the earth
Drop out of air.

Red stars of light
Burn on the round
Of land: street-constellations
Strew the ground.

Their bridges leap
From town to town:
Into lighted dusk
We circle down.

Still as a bird
Transfixed in flight
We come to nest
In the field of night.

James Kirkup

In Memory of James Eagleson, RCAF

I

In what glorious air
an acrobat he grew,
sailing circus boy
in his net of air,
confident,
setting the Mitchell always down
and running out for one
last bow. Roar
of engine and love a limit
he knew, he knew so much
the crew, no trapeze
was that high, they could not
swing laughing and tense
Berlin, Hamburg, Bremen,
a tangled run to Ruhr
and back to pile out like apes
swinging from harness into
the happy jungle of their own.
Face in a NAAFI mug
and letters from home.

II

Trip of hammer and hell
black eruption of sky
and shell in September sun
and a rattling chance to go in
low and come out
like swimmers.

In the eyes quickly closed,
in the captured light, there came
sudden heat of red and white
filling the cabins of their skulls
with its close warmth
dull and dry.

The left motor hit square.
And in that minute
of war and witness
they were headed down.

Douglas Lochhead

Airborne over Orkney

Outstretched below
the isles of Orkney lie:
skins of ancient monsters
patterned wildly.
Long sinuous members, golden fringed,
sleep deeply, unstirring;
great nostrils, quiet now
lapping a tide.
Abrupt headlands: lurking claws
mindful of lost sea struggles.

The sea, lightly creased,
caressing her trophy,
sings of the vanquished.

Christine De Luca

Epitaph for an Air Crew

This is a thought to sain and save –
They did not grudge the gift they gave.
This be for joy and this for pride –
That other men might live they died.

Daedalus

He made a mouse trap
and his PRO man
called it a labyrinth.

It was to catch a mouse –
the local papers
called it a Minotaur.

In the film they made of it
Theseus and Ariadne
got bit parts.

But the spectators loved them
(projections of themselves)
and called them stars.

Then he made his mistake.
He forced his son
to follow in daddy's wingbeats.

Icarus rebelled and fell
through the generation gap
into mythology.

In his way he won – his name
is on the map,
which is more than his father's is.

Norman MacCaig

JET

Pushes sound before it
   until the sharp snout splits
   the membrane of thin air

and it leaps ahead of expectation
   screams to be free and gone
   gone from now
   and off that beaten track of time
it shares resentfully with you and me.

And the world and all its old wild things
   must cower and bow down
   beneath that shriek
of agony and power
   in the torn sky.

Bill McCorkindale

ICARUS

The cloud sea drifts into the morning sky.
Under your wing I linger, would remain
safe in the aqua fading into blue.
Soft clouds below obscure the sodden earth
the mountain's dark insistence, and the sprawl
of cities forced into the light of day
out of the night's warm slumber.

                                    Here with you
curled in your arms, love-drowsed, I do not hear
the break of day, the passing of our dawn.
the morning's dappled call. I can ignore
the sun that plays upon your golden wings,
warm now and soft; too late I wake and turn
shocked by the sudden heat, the burning wax –
surely we will not fall?

Anne MacLeod

THE FUTURIST AVIATOR SPEAKS
TO HIS FATHER, VULCAN

I come to you, Vulcan, to give back the laugh
to you, sputtering, old ventriloquist.
Believe me, I'm out of your reach!
You'd snare me if you could,
in your coils of lava,
that luck you have with foolish dreamers
who climb your slopes
when the hypnotizing sadness of your monolithic sunsets
convulses into horrid, titanic guffaws,
and sometimes an earthquake.
I fear neither omens, nor menace of the abyss
that at your whim can bury a city
beneath a tumulus of ore and ash and blood.
I am the Futurist, strong and indomitable,
hauling aloft my wild and enduring heart:
and so it is I set me down at Aurora's board,

and feast upon her color-show of fruits;
or trample meridians, launch my bombs,
pursue the fleeing armies of the sunset,
dragging the wistful, sighing twilight
in tow behind me.

Etna, Etna, who dances better than I
pirouetting above your fearsome maw
bellowing a thousand meters below?
Watch me descend and dip toward your sulphurous breath
and dart between your columns of reddening clouds
to listen to the rumbling of that vast belly,
your heaving, gulping, deafening landslide,
Your war at the centre of the earth.
In vain your carbon rage
that would buffet me back to the sky!
I grip the flight-stick firmly in my hands...

I enter now, through the wide gap of your mouth,
a sprawl of peaks,
and drop still further down
to inspect your monstrous gums...
Vulcan! what weeds are these
limp plumes of smoke
you nibble at,
like an ogre's blue moustache?....

Filippo Tommaso Marinetti
translated by Felix Stefanile

A PARACHUTE PACKER

my calloused hands
stumbling
gained sureness
for lives depended
on my folds
over and over
the scallops of sensuous silk
spinning dreams
in the tedious housework of war
like dishes never done except
if you leave a bit of food
in the crevice of a plate
so what perhaps a stomach ache
and the girls at the party last night
ah yes
folding parties
into parachutes
silk panties
into parachutes
and lives
depended on my folds

Dona Paul Massel

THE FLYING VISITOR

Do you believe in the fifth dimension ?
Of spectre, spirits, phantoms and ghosts,
On an airfield called East Fortune,
An apparition, who is mine host.

Many say that they have seen him,
Felt his presence, icy air,
Turned around to try and meet him,
Is that someone standing there ?

A ghostly shape clad from the past,
Leather coat tattered with wear,
Face is tainted from exhaust soot,
Leather gloves in disrepair,

Steals around this old airfield,
From hangars to outlying sheds,
Making the odd timed manifestation,
Watching old haunts, it's been said.

Now this ageless flying fighter,
A forgotten hero it could be said,
Silver wings mounted on his chest,
Hat and crest upon his head.

Will he ever find his freedom ?
What on earth holds him here ?
Does he have a ghostly secret ?
We will never know …. I fear!

Peter M McCulloch

The Harvard

Over blue skies at East Fortune
Reels a warrior cammo green,
A defender strong and fearless
A surviving fighting machine.

This trace of cherished history
Draws air deep in its heart,
As down to earth it plummets
Tearing atmosphere apart.

Swooping beyond an ecstatic crowd
Synchronising every eye,
All ages standing awe struck
As this performer screws the sky.

Banking over tree and field
A burble from the east,
Drawing nearer to its mark
His air speed now increased.

Skirting round the hangar walls
An emerging blur of green,
Starboard wing tip inviting earth
Or gravity to intervene.

In a flash he soars so high
A spellbound crowd in a trance,
Banks again and levels out
I can sense a flying romance.

But now this proud performer
Bids farewell on tilted wing,
As we applaud his show of courage
And memories...
Where do you begin?

Peter M McCulloch

AIRSHIP

Recovered from pale blueprints
and forgiven its heritage of charred metal
the airship moves at the wind's direction
through the next world. A high
slipstream of time
brings it in view: just
bouncing, it seems, from cloud-edge
to treetop, almost a milky bubble.

Now, this moment we peer,
throats tensed ready to shout,
the ship tilts its nose to the sun
and its oval shadow contracts to a grasspatch
as it shimmers and disappears.

What message arrives from the mariner
trapped in this bottle? Silence.
A freak technology has lifted his tongue –
someone, somewhere, knows and speaks his name:
perhaps he's among us now, not yet alone.

Roger McDonald

HIGH FLIGHT

Written by a nineteen year old RCAF pilot who was killed
in action on 11th December 1941.

Oh, I have slipped the surly bonds of Earth
And danced the skies on laughter-silvered wings;
Sunward I've climbed and joined the tumbling mirth
Of sun-split clouds – and done a hundred things
You have not dreamed of – wheeled and soared and swung
High in the sunlit silence. Hov'ring there,
I've chased the shouting wind along, and flung
My eager craft through footless halls of air ...

Up, up the long, delirious burning blue
I've topped the windswept heights with easy grace
Where never lark, or even eagle flew –
And, while with silent lifting mind I've trod
The high untrespassed sanctity of space,
Put out my hand and touched the face of God.

John Gillespie Magee, Jr

HOMERIC SIMILE

As when a heavy bomber in the cloud
Having made some minutes good an unknown track;
Although the dead-reckoner triangulates
Departure and the stations he can fix,
Counting the thinness of the chilly air,
The winds aloft, the readings of the clocks;
And the radarman sees the green snakes dance
Continually before him in attest
That the hostile sought terrain runs on below;

46

And although the phantom shapes of friendly planes
Flit on the screen and sometimes through the cloud
Where the pilot squints against the forward glass,
Seeing reflected phosphorescent dials
And his own anxious face in all command;
And each man thinks of some unlikely love,
Hitherto his; and issues drop away
Like jettisoned bombs, and all is personal fog;
Then, hope aside and hunger all at large
For certainty what war is, foe is, where America;
Then, the four engines droning like a sorrow,
Clear, sudden miracle: cloud breaks,
Tatter of cloud passes, there ahead,
Beside, above, friends in the desperate sky;
And below burns like all fire the target town,
A delicate red chart of squares, abstract
And jewelled, from which rise lazy tracers,
And the searchlights through smoke tumble up
To a lovely apex on some undone friend;
As in this fierce discovery is something found
More than release from waiting or of bombs,
Greater than all the Germanies of hate,
Some penetration of the overcast
We make through, hour upon uncounted hour,
All this life, fuel low, instruments all tumbled,
And uncrewed.
        Not otherwise the closing notes disclose,
As the calm, intelligent strings do their duty,
The hard objective of a quartet, reached
After uncertain passage, through form observed,
And at a risk no particle diminished.

William Meredith

THE WOODCOCK

The jet plane flew
As low as it dared -
Tearing the air
With flame and roar.

The woodcock flew
As low as it could -
Hugging the ground
Without a sound.

John Morrison

TO A BARRAGE BALLOON

We used to say "If pigs could fly!"
    And now they do.
I saw one sailing in the sky
Some thousand feet above his sty,
    A fat one, too!
I scarcely could believe my eyes,
So just imagine my surprise
To see so corpulent a pig
Inconsequently dance a jig
    Upon a cloud.
And, when elated by the show
I clapped my hands and called "Bravo!"
    He turned and bowed.
Then, all at once, he seemed to flop
And dived behind a chimney-top
    Out of my sight.
"He's down" thought I; but not at all,
'Twas only pride that had the fall:

To my delight
He rose, quite gay and debonair,
Resolved to go on dancing there
Both day and night.

So pigs can fly,
They really do,
This chap, though anchored in the slime,
Could reach an altitude sublime –
A pig, 'tis true!
I wish I knew
Just how not only pigs but men
Might rise to nobler heights again
Right in the blue
And start anew!

May Morton

PRESTWICK AIRPORT

Here the world's great walked on our common ground,
though we had history before they came:
Wallace once stood upon a nearby mound
to watch the well-stocked barns of Ayr aflame.
When I was young they called it Orangefield:
Ball and McCudden used to fly from here,
flat western farmland of the fogless bield
long before radar made dark heaven clear.
Now to new fields the flying galleons sail,
tracing their glide-paths over city walls.
Where once the Sleeping Warrior marked the trail
the ghosts of queuing travellers haunt the halls.

But here among the phantoms and the blues
Elvis touched Scotland once in G.I. shoes.

William Neill

BOMBING RUN, EASTER ROSS

In quick succession, two jets tear
slanting down and out to Whiteness Sands.
Skull-pounding thunder echoes where they were
short seconds back, like clapped atomic hands.

on empty sky. From along the bay
follow smaller sounds, these no less
threatening; six quick whip-cracks detonate
efficiently as a building site blast

through earth. They seem too vulnerable
in that huge wide blue, too lightweight
to matter, but such power, capable
of taking out entire lands, states.

No-one here seems to notice. If caught
in mid-conversation, they'll simply stop
until there's silence, still as rabbits
in the face of overwhelming odds,

then carry on as if the world's drawn
back together. Where once was continuous
sound and sense from the sea, now comes
the crashing sound of its absence.

I freeze while walking round the north's long coast
at the booming surf of the future
taking out, heartbreakingly, the past
with precision's perfect fracture.

Stuart A Paterson

TINA'S FLIGHT

Some roads look as if they might go on forever
The way they twist and weave from place to place,
And others seem not really roads at all,
But runways sliced off from Earth and built
To launch us up, and off to Heaven.
I last saw you on such a road, your direction
Narrowed to harrowing certainties.
Head down, eyes peering from under a blonde fringe,
Arms held out steady like wings,
Your short future long diagnosed
You prepared to speed towards the sliced-off horizon,
And launch yourself Heavenward.
Gone now, you've left behind
A slowly dispersing trail of years.
I raise a glass of wine to a summer cloud,
To a child's balloon on its maiden voyage,
To your last signature scrawled on the sky's
Far-off, damson-coloured nothingness.

Brian Patten

THROTTLING THE PILOT

Abrasive altitude fog-snakes Heathrow,
such scarlet runway-lights that define speed
are occluded, intermittent, then dead.
Klaxons patrol the vertigo
maelstromic in the cumulose sinews
of siamese-beige smog. The joystick threads
the turbines to their maximum circuit.
Nausea balloons its terminal green threat.

And one detached from his reservation,
operative on the transfusion of fear,
gun-walks his balance of a saboteur
in contracted bourbon-features.
An instant's annihilative future
determines his detonative facade;
green airline handbag armed with precision-
silencers. The turbines throttle back speed,

circumnavigating the control-tower.
Quavering in sense-querulous turmoil
the passengers vibrate in maroon-cells
paralysed by percussive height.
Hysterical to unseal a flight-door
through suicide victimize what's rational,
as height accelerates beyond the last
red signal, orbiting devoid of fuel.

Jeremy Reed

FLY

Glide, soar, more and more
high in the sky
up, up, rise
then dive
down, around
windblowing, soaring
over hills, above the trees
sweep and swirl
fall
then surge
climbing, ascending
fly like a bird
free, at ease
ride the breeze
hover
sit in the air
still, calm, devoid of noise
looking down
scan around
seeing near and far
corn, fanning back and forth
moors, woods, wilderness
upwards
river winding, finding sea
mountain peak
now descending through the cloud
further still
hill and fell
landing gear down

lower still
tree top height
slowing down
level ground
on a perfect night
end of flight

James Reid

BALLAD OF A BALLOONIST

The blue polyester is already pregnant,
displaying some sponsor's name.

In this thickening dusk
there's the quiet roar of a gas-flame.

The passenger takes nobody with him.
His hands shine on the wicker edge.

This morning was normal.
It isn't that he doesn't like it here.

Such solitary flights can't be so bad
except that you might get lost,

never heard from again.
He's dumping sand-bags like dead pigs.

The stiff ropes untied,
we offer honey-water and milk.

If the takeoff gives him a kick
he's careful not to show it.

Winds may drive him where no one guessed,
blue tigers leaping from nowhere.

He rises at an acute or obtuse angle,
gliding as an angel along the lawn.

Though he controls his altitude
he can't choose to turn left or right.

It's difficult to discern him now,
tossed by cross-currents.

His balloon charges like a blue cow
through the universe.

We're told he hopes to descend
on the other side of tonight.

Susanna Roxman

REICHERT'S LEAP

In December 1911 Walter Reichert, a self-employed tailor,
attempted to fly from the Eiffel Tower.

Each hopeless stitch homemade - the futile skin
Out of which he would jump. It's too late now,
Even if he wanted, as surely he must be wanting,
To climb down. He crouches on the cold brow
Of madness, on the parapet's fine brink,
His breath smearing the air, an unused silence
Which might have been saving speech. *Here, what do you think?*
I didn't mean it. Of course. It makes no sense.
Or simply, *We'll have a drink and then I'll go*
To embrace my poor wife and children. Later, we'll chat.
Too late. They say the journalists wouldn't throw
Their story out of the window, and that was that.
He shifts like a bird – ridiculous, he fears.

Go on, you're chicken. He cannot, will not stay
For this. A puff of looping breath. The sheer
Drop. The stupid tower begins to sway
In his mind only. A final shuffle and
A plop as into water. The deficient air
Fails to support him. A black duster lands,
Bundles into the ground. A brief affair.
Men in caps. Fuss. A canvas shroud.
A way elbowed through the encroaching crowd.

Which of us, from our tower, would not recall
This brute parabola of pride and fall? –
Late Romantics, fledgling birdmen all.

Lawrence Sail

FLIGHT 000

It sags, the bright machine,
From level to level, swooping
Down beneath the canopy
Not of sky but forest.

Not clouds, but green boughs
Flick and snap at the windows,
Grazing the glass like rain
On through the avenue of trees.

And all this time, vainly,
The meticulous self-adjustments,
The throttling up and down,
The engines' whining obedience...

And this is the dream's substance –
Not the crash that must come,
Not the brute brunt in which
Limbs are a scattered luggage –

But this: the silver body
Perfect, graciously tilting
Along the narrowing flightpath.
Its illusory odyssey.

Lawrence Sail

... Jet

Nor am I all that keen on this
    Enormous man-made bird,
And 'Jumbo' doesn't seem to me
    At all the proper word.

I've never seen an elephant
    Stretch out great shining wings;
Though, come to think of it, I've never known
    An aeroplane that sings.

But 'bird' does seem a better word
    Than 'Jumbo' to declare
The shape and the behaviour of
    This giant of the air.

They tell me flying is quite safe –
    Safer, so they say,
Than driving in a coach or car
    Along the motorway.

But I recall when Grandpa said,
   That what he'd always found
Was wise folk keep a level head
   And both feet on the ground.

Vernon Scannell

FLIGHT OF BALLOONS

They rise, enormous bubbles, from the ground,
Transform familiar sky with fantasy;
Ascension leaves behind terrestrial sound;
They float on silence, mocking gravity.

Reduced by distance, all the earth beneath
Becomes a children's game for rolling dice –
Green shaded patterns, meadow, copse and heath;
The roads, thin tapes, and hillocks, sleeping mice.

And we, below, gaze up and see the bright
Inverted garlic bulbs suspended there –
Historic images of early flight –
On their transparent threads of light and air.

How little they have changed since man first rose,
Two centuries ago, in one like those.

Vernon Scannell

WING AND ARROW

How strange and diverse is man !
The girl who wears a badge of wings
Holds hands, across tea cups, with the man
Who breaks the bomber's flight with rushing shell.
Thus does the bird love the arrow,
And thus the bow kiss the wing it pierces.
All hold hands to love and kill
Because they wish to kill and find it sweet to love.

Richard Spender

BEFORE THE FIRST PARACHUTE DESCENT

All my world has suddenly gone quiet
Like a railway carriage as it draws into a station;
Conversation fails, laughter dies,
And the turning of pages and the striking of matches cease.
All life is lapsed into nervous consciousness,
Frozen, like blades of grass in blocks of ice,
Except where one small persistent voice in the corner
Compares with the questioning silence –
With the situation of an electric present –
My self-opinions, pride and confidence of an untried past.

Richard Spender

PLANES

That close the jet is
Ye micht see the mannie
I' his wee houss o gless,
Some unkent gladiator,
Him wi his glintin helm.

And ablow me   *[below]*
And ablow him
The shadda o his plane
Is crucifie't be the sun
Whyle it races ower the strand   *[stretches]*
O gerss and heather.   *[grass]*
Syne he airts the tips o his wings   *[directs]*
Atwein yird and space   *[between earth]*
And wheiks round the heid o the glen.   *[whips]*
The hauf-orphan't roar that belangs
The fire i' his tail
Crashes and stachers eftir him throwe the hills *[staggers]*
And he is awa.

The auld propellor plane
Toils ower the lift like a clock   *[sky] [beetle]*
And like the clocks ye watch
Skutterin ower a stane,   *[scurrying]*
He is heidit wi a purpose
Secret til hissel.

Gavin Sprott

60

SKYDIVERS

They fall outwards
as if from the calyx of a flower
each smaller than a falcon's claw,
their target a gravel circle
in the Byzantine barley.

They fall like hushed flame
where once the sun's disk
was ploughed from the furrow,
coupling, uncoupling
above the drop-zone.

When they run
with the white squall
you would think the air
holds their flight
like a welder's seam,

but as they alight
there's a sudden
billow of pollen,
an uprush
from winged heels

and like lovers,
the sweet tarrying
of their bodies
dissolves
the moment.

Pauline Stainer

HUBRIS, OFF THE WHITE CLIFFS

Our bomber flying home
drops eight remaining bombs.
To see the water plume
she circles round the bombs.

For spectacle and noise –
REVERBERATING BOMBS –
she circles in the noise
and passes round the bombs.

A flight of Messerschmitt,
remembering the bombs,
fires at the plane and it,
in smithereens, like bombs

drops and makes another pass
across the pluming bombs.
The Messerschmitt then pass,
then leave the pluming bombs.

Martin Stokes

IN PRAISE OF FLYING

Low Concorde
    coming in to land.

Silent leaf
    drifting on the wind.

Daniel Stolfi

AIRPORT

Palace of unreality, where the place
we have just been to fades from the mind – shrinking
to some scribbled accounts, postcards unmailed,
and faces held dear, let go, and now sinking
like coins in clouded, forgetful water –
and the place we are heading toward hangs forestalled
in the stretched and colorless corridors,
on the travelling belts, and with the false-

smiling announcements that melt in mid-air:
to think, this may be our last reality.
Dim alcoves hold bars well-patronized but where
there is not that seethe of mating, each he and she
focused instead on a single survival.
To pass through, without panic: that is all.

John Updike

OTHER TITLES FROM NMS PUBLISHING

POETRY

Translated Kingdoms – poems of
Scotland and the sea
Present Poets
Scotland and the World

SCOTLAND'S PAST IN ACTION SERIES

Fishing & Whaling
Sporting Scotland
Farming
Spinning & Weaving
Building Railways
Making Cars
Leaving Scotland
Feeding Scotland
Going to School
Going to Church
Scots in Sickness & Health
Going on Holiday
Going to Bed
Shipbuilding
Scottish Bicycles & Tricycles
Scotland's Inland Waterways

Forthcoming titles:
Engineering
Getting Married
Scottish Music Hall
Cinema in Scotland
Brewing

SCOTS LIVES SERIES

The Gentle Lochiel
Elsie Inglis
Miss Cranston
Mungo Park

Forthcoming titles:
Scottish Suffragettes
Scotland & Slavery

ANTHOLOGY SERIES

Treasure Islands
Scotland's Weather
Scottish Endings
The Thistle at War

ARCHIVE PHOTOGRAPHY SERIES

Bairns
Into the Foreground
To See Oursels

GENERAL

Scottish Coins
Tartan
Scenery of Scotland
Viking-age Gold & Silver of Scotland
The Scottish Home

Obtainable from all good bookshops or direct from NMS Publishing
Limited, Royal Museum, Chambers Street, Edinburgh EH1 1JF.